A record block of granite at Polkanuggo Quarry, Cornwall, in September 1902. 110 pounds (50 kg) of black powder were used to dislodge this block, estimated to weigh 2,738 tons.

QUARRIES AND QUARRYING

Peter Stanier

Shire Publications Ltd

CONTENTS

 First published 1985. Shire Album 134. ISBN 0 85263 728 4.

British Library Cataloguing in Publication data available.

Set in 9 point Times roman and printed in Great Britain by C. I. Thomas & Sons (Haverfordwest) Ltd, Press Buildings, Merlins Bridge, Haverfordwest, Dyfed.

ACKNOWLEDGEMENTS

The author is grateful for the help received from Christopher Hall, David Pollard, Roger Penhallurick, Ian Rutherford and the Burlington Slate Company Ltd. Photographs on the following pages are acknowledged to: the British Geological Survey (NERC copyright reserved), 7, 11 (lower), 15, 18, 29; Gwynedd Archives Service, 24; C. J. Hall, 11 (upper); Sam Hanna, 31 (lower); Llechwedd Slate Mines, 27 (lower); Old Delabole Slate Company, 26, 27 (upper); A. E. McR. Pearce, 10; Mr and Mrs C. Piper, 16 (upper); the Royal Institution of Cornwall, 3, 17, 22, 23, 25, 28 (upper); Stothert and Pitt plc, 30; H. Tempest (Cardiff) Ltd, 28 (lower); Weymouth and Portland Museums, 6. The remainder, including the cover photograph, are from the author's collection.

COVER: *Weston Quarry, Portland.*

Several firms specialised in manufacturing quarrying and stone-working equipment, particularly in Yorkshire. This advertisement dates from 1905.

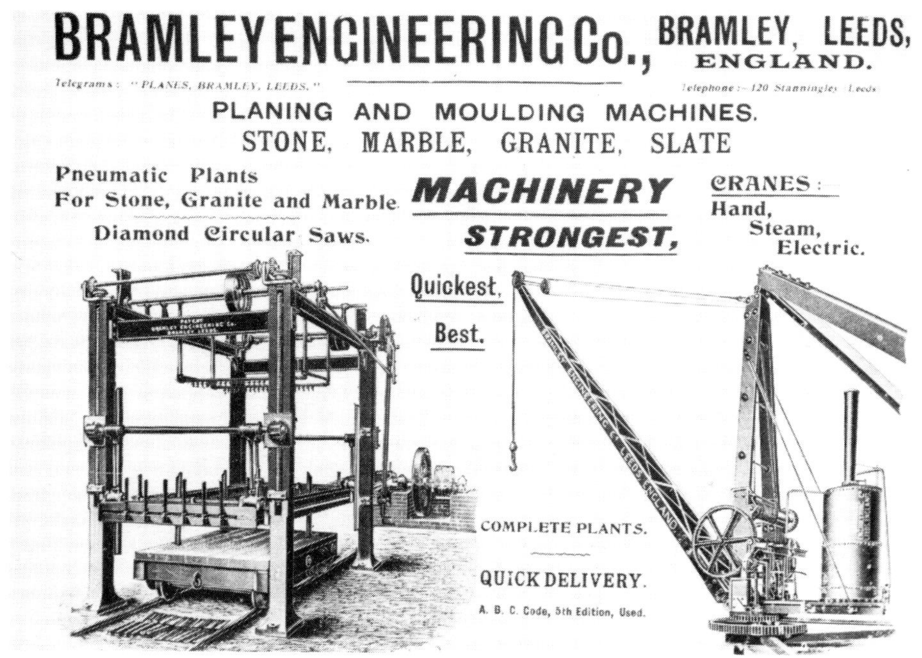

Traction engine 'Beta' loading kerb stones on to a wagon at a granite quarry near Penryn, Cornwall, in the late nineteenth century. Note the timber derrick crane of the period, with a masoned stone held by a chain and special hooks called 'dogs'. Quarry tools were sharpened in the smithy in the background.

QUARRYING STONE

Stone quarrying has been a widespread and important industry in Britain. There are thousands of workings, ranging from humble overgrown pits which once supplied local needs, to large, highly organised quarries, exporting their products to many parts of Britain and even abroad. The word 'quarry' is derived from the Low Latin *quadraria,* meaning 'places where squared stones were cut'. This book examines mainly those quarries where quality dimension stone has been sought for architectural, engineering or monumental work. A geological map will quickly show the great variety of rocks found in Britain. In general, the older igneous and metamorphic rocks are located in the upland north and west, with the softer sedimentary freestones in the south and east.

All rock types have been exploited at least locally at some time in the past. The Romans were probably the first true quarriers and workings survive, for example, close to Hadrian's Wall and at Chester. Medieval quarries were larger, with much activity along the outcrop of Jurassic rocks from Yorkshire to Dorset, especially during the period of cathedral building. The famous quarries at Barnack (Cambridgeshire) consisted of a wide area of surface pits. Quarrying was not on a large scale until the eighteenth and nineteenth centuries, when there was increasing demand from large towns and industrial districts. In most cases, the output of quality stone reached a peak at the end of the nineteenth century, after

Compressed air plant became increasingly important at quarries in the twentieth century. These three Broom and Wade compressors were driven by Gilkes turbines (right) and were installed at De Lank Quarry, Cornwall, in 1927.

which decline was often rapid in the face of competition from other countries and the cheaper materials of the twentieth century. Where there were once hundreds of quarries, there are now only a handful, with a scarcity of skilled workers. However, the decline may have halted in the 1980s, with an increased demand for natural stone products.

Only the better-quality stone could profitably be carried far, and the development of successful quarrying has always depended on good transport. Sea transport was the easiest method, as was realised by the Normans, who imported Caen stone from Normandy to England. In the seventeenth century Sir Christopher Wren recognised the convenience of Portland's coastal site for obtaining stone for rebuilding London after its devastation by the Great Fire. For all the rock types described here, ready access to the sea has been of prime importance, while rivers, canals and railways aided development elsewhere, especially during the nineteenth century.

The Quarries Act of 1894 defined a quarry as a working for stone which was over 20 feet (6 m) deep, although in some cases surface rocks were also taken. The ideal quarry is cut into a hill or valley side, with a level entrance. Geological or other conditions may prevent this and an open pit might be required, where cranes and pumps would be essential. Where good rock dips deeply, quarries have been continued underground at many places. Although such workings employ quarrying techniques, the term *mine* is used in this book to make the distinction. A well laid-out quarry of about 1900 would include cranes, tramways, dressing sheds and a blacksmith's shop for making and sharpening tools. Although water was used where available, other sources of power became increasingly important. Steam and compressed air took over from hand-power for the largest cranes and were soon applied to rock drills. Steam and oil engines provided power for driving air compressors and dressing machinery. Electricity is now widely used for the

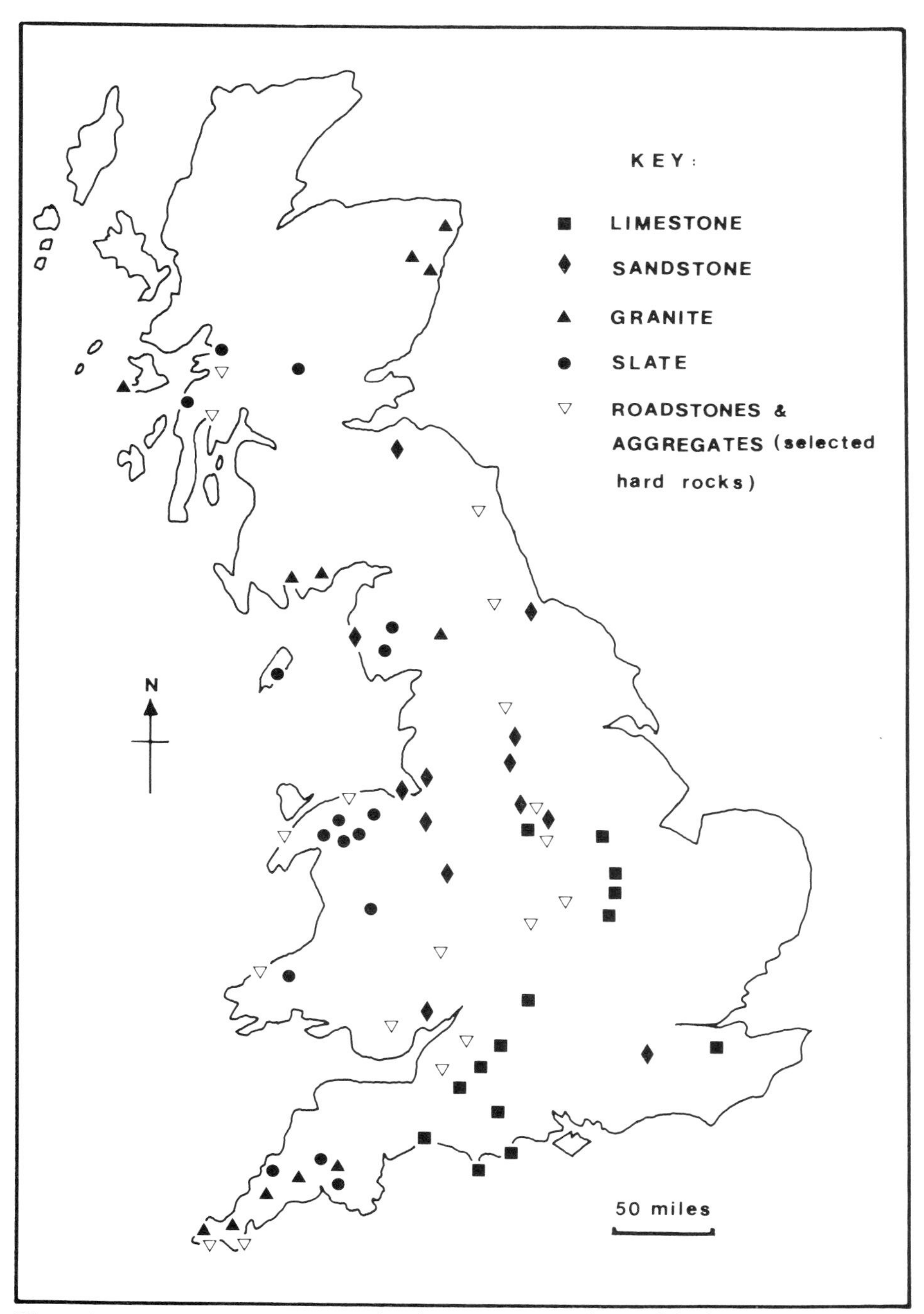

The main quarrying areas in Britain for dimension stone, slate and broken stone.

latter. If explosives were used, a solid-walled magazine would be sited some distance away. As the industry took only the best stone, the dumping of much waste has resulted, except where alternative uses for it were found, such as for roadstone.

Conditions for quarrymen and stone masons were often hard. They worked outdoors with little shelter for much of the year and were commonly paid only for usable stone quarried or dressed, but not for waste or breakages, even if these were due to an unseen flaw in the stone. The more common accidents, sometimes fatal, were caused by falling rocks, blasting, cranes or tramways, and eye injuries might be incurred while dressing stone. There was also the long-term danger of silicosis, from inhaling dust, especially while working granites and sandstones.

Transporting Portland stone in 1805, using a two-wheeled 'jack'. Note the horses at the rear acting as the braking system. Conditions were improved by the Merchants Railway of 1826 and, at a later date, the introduction of traction engines.

Building a 'bench' at Suckthumb Quarry, Portland. This was the usual method of disposing of waste overburden rock as the face advanced. The Portland stone is found at the bottom of the narrow gully behind the crane. By 1930, steam cranes had replaced many of the timber derricks, once a feature of Portland.

LIMESTONE

Limestone is a common rock type, calcium-rich and often containing much broken shell materials. It was originally laid down under the sea in sedimentary beds, separated by horizontal bedding planes, which are now lines of weakness along which a bed can be detached from a quarry. Vertical joints or faults, formed by subsequent earth movements, aid the extraction of rectangular blocks. Where they are widely spaced, wedges or saws have been used to obtain smaller blocks (explosives would damage quality stone). The softer, finely grained and easily worked limestones are known as freestones. The valuable Jurassic limestones are mainly oolites, with tiny rounded grains and shell fragments cemented together, and are noted for their attractive colouring and ease of working. They were an important source of building stone in medieval times, and notable quarries were at Ham Hill and Doulting (Somerset), Weldon (Northamptonshire), Ketton and Clipsham (Leicestershire). Throughout the belt, there are differences in the geology of the stone and the methods of extraction.

The surface of the Isle of Portland (Dorset) is scarred by quarrying. Portland stone is famous for its use in St Paul's Cathedral, London, and elsewhere, although it had been worked for centuries before. The first workings were along the cliffs and landslips in the higher northern part of Portland but, as this source became exhausted, quarrying moved inland. Transport to shipping points was greatly eased in 1826, when the Merchants' Railway connected the quarries via an incline to Castletown Pier. In the south there were small quarries on the cliff edge, from which cranes loaded the stone directly into barges. By 1839

A compressed air derrick at Gregory's Quarry, Ancaster, in 1984. Note that the waste stone is dealt with in a similar way to Portland. The soft overburden is clearly visible above the limestone face.

there were fifty-six different quarries, many of them family concerns. Convict labour was soon used in government quarries for major works such as the Portland Breakwater (1849-72). In 1865 Portland was linked to the rest of England by railway, and over 80,000 tons of stone were carried in its first year. Since then the industry has flourished and declined. The last major work was the supply of stone for the reconstruction of London after the Second World War. Only a few small quarries remain in operation.

Inland, it was first necessary to remove up to 60 feet (18 m) of overlying rocks, of which the tough *Skull Cap* was broken up by wedges or blasting. Where possible, this waste was tipped over the cliffs, but elsewhere it was carefully dumped within the quarry as the face advanced. The beds sought were the *Roach,* a hard shelly limestone good for heavy engineering, the *Whitbed,* the true white oolite up to 10 feet (3 m) thick, and, below this, the softer *Curf,* used for interior work. Huge blocks were obtained by using wedges in the joints, but while the Roach was still attached to the Whitbed the whole block was hauled on to its side by cranes (earlier, by jacks) to allow a V-shaped groove to be cut at the junction of the two beds. Into this were placed iron wedges between thin plates, and these were driven home with sledge hammers until the beds separated. A similar method for reducing the size of a block was by *pitting* a shallow groove, into which steel wedges and *scales* (plates) were placed as before. Blocks of up to 10 tons or more were squared with a *kivel,* a tool consisting of an axe at one end and a hammer, slightly hollowed to give two cutting edges, at the other. This traditional tool was described by John Smeaton when he visited Portland in 1756, while selecting stone for the interior of his Eddystone Lighthouse. Channelling machines and saws have been used for extracting, but today the rock is split with

ABOVE: *Drilling a line of holes at Gregory's Quarry, Ancaster. Plugs and feathers are inserted to detach a long slab of rock. Note bedding planes and marks of previous holes in the face behind.*
BELOW: *Quarried stones with painted stock numbers and carved Roman numerals indicating dimensions in cubic feet. Many quarry districts have their own tradition of markings; these are at Ancaster.*

Underground at Monks Park, Wiltshire, in the 1930s. The bottom bed of Bath stone is being sawn out with 'frig bobs' by the light of oil lamps. Water was fed into saw cuts as a lubricant. Note the pick being used to clean the stone surface.

plugs and feathers (see below), while circular saws are used in the finishing process.

200 miles (320 km) away at Ancaster (Lincolnshire) the limestone is softer and more shelly. This inland source of stone has been exploited since medieval times for buildings as distant as Cambridge and London. Gregory's Quarry is still worked in the traditional way by only four men. A soft clayey overburden is easily removed to reveal the *Weatherbed* (shelly) and *Hard White* beds. A compressed air drill is used to make a line of vertical holes about 2 feet (610 mm) deep. Into each hole large steel *plugs* (tapering wedges) are inserted between two *feathers* (thin plates) and driven home evenly with a sledge hammer. This splits the rock along the line, while it becomes fully detached along a horizontal bedding plane. The block is reduced to manageable sizes by a similar method before being lifted out by crane.

Where it was uneconomical to remove the overburden, stone beds were worked underground, notably around Bath and Corsham, which have the best known mines (called quarries locally). They have been used since Roman times, but the greatest exploitation began in the eighteenth century, largely owing to Ralph Allen, who built Prior Park in 1735, partly to advertise the quality of the stone from nearby Combe Down. A tramroad of 1731 took stone to the river Avon, whence it was shipped donwstream to Bristol. After 1810 the Kennet and Avon Canal opened up the possibility of trade towards London. The greatest source, however, was a few miles east at Box, where extensive beds were discovered during the building of Brunel's Box Tunnel for the Great Western Railway (1836-41). The resulting huge stone mine had a direct rail connection and by 1864 some 100,000 tons were sent every year from Corsham station to all parts of

RIGHT: *Lifting out a sawn block of Bath stone at Monks Park. This short crane is made of stout timber and is fixed between floor and ceiling but differs from the usual underground cranes in that the jib appears to be adjustable. The hook is attached to a 'lewis', locked in a tapering hole cut in the block. Other lifting methods included 'shears' (like calipers), gripping the sides of the stone. Note the pillar of uncut rock beyond, left to support the roof.*

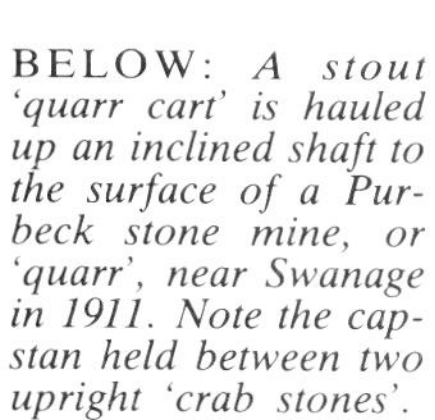

BELOW: *A stout 'quarr cart' is hauled up an inclined shaft to the surface of a Purbeck stone mine, or 'quarr', near Swanage in 1911. Note the capstan held between two upright 'crab stones'.*

ABOVE: *A reciprocating stone saw at Lawn Quarry, Combe Down, Bath. Circular saws are more common at other quarries and stone yards.*

BELOW: *As with other freestones, Combe Down stone can be worked easily with hand tools. This is part of a lantern cap for Brighton Pavilion, being worked at Lawn Quarry in 1984.*

Britain. Other mines were opened in the Corsham area, with access by slope shafts from the surface. The workings extend for many miles.

Stone was first extracted with picks and *jadding irons* (long rods), but these were replaced by large saws up to 7 feet (2 m) long. A pick was used to make a space between the top of the bed and the roof above, so that a narrow *razzer* could be inserted to start the first cut. The larger *frig bob* finished the job and once the vertical cuts were made *wedges* and *chips* were driven in to detach large blocks along the natural horizontal bedding. A trapezoidal *lewis* was inserted into a tapering hole cut in the block to enable it to be lifted out on to a tramway wagon. The short timber hand cranes were fixed between the floor and a square *chog hole* cut in the roof. After the Second World War hand saws were replaced by mechanical Samson cutters. Wide pillars of uncut stone have always been left to support the roof. The stone is filled with *quarry sap* (natural fluids) and must be allowed to dry out for at least six months in some cases, being stored underground in winter to avoid frost damage. In the inter-war years many firms in this area and at Portland were amalgamated into the Bath and Portland Group. Some mines were taken over by the Government. Most are now abandoned, along with lesser known mines further north in the Cotswolds. Only two mines remain at work in west Wiltshire.

Elsewhere in Wiltshire, there were stone mines at Chilmark, from which came much of the stone for Salisbury Cathedral. Rock drills were employed here in the 1930s, when mining finished. A pale chalk rock, turning darker and harder with exposure, was worked underground at Beer (Devon) and was used in the interior of Exeter Cathedral. The earlier chambers were vaulted but when they were reworked with saws in the 1900s by Wiltshire quarrymen rectangular pillars were left. In Dorset there were underground workings along the cliffs west of Swanage, while a little inland were stone mines, called *quarrs,* for Purbeck marble, a freshwater shelly limestone used for polished interior work. Other hard limestones which could take a good polish have been worked locally as 'marbles', such as at Ashburton (Devon) and the Ashford black marble of Derbyshire. The crinoidal Hopton Wood limestone, also from Derbyshire, is famous for its decorative properties.

SANDSTONE

Sandstone is another sedimentary rock, laid down under marine or desert conditions. It contains mainly quartz particles, cemented by silica or other minerals, giving a great range of colours. When fine-grained and easily tooled it is a freestone. Sandstones are well bedded and have been worked by similar methods to the Jurassic limestones. As with limestones, they are widespread and of variable quality. The Carboniferous sandstones are the strongest and include the coarser-grained millstone grit, mainly in Derbyshire and the Yorkshire Pennines. When fresh, it is a golden brown or grey. It was important for millstones and for grinding, especially for the cutlery industry of Sheffield, with quarries along the Millstone Edge and other edges near Hathersage (Derbyshire). There were quarries for building stone at Darley Dale, near Matlock, while in West Yorkshire quarries in the Bradford and Halifax areas have been important, sending their products to other parts of Britain.

In Scotland the whitish grey Craigleith stone was widely used in Edinburgh and shipped to London. Pennant sandstone was important in South Wales and around Bristol. At Mansfield (Nottinghamshire) white and red stones have been quarried for many centuries. New Red Sandstones (Permian and Triassic) have been important in the Shropshire and Cheshire area, including quarries at Grinshill (Shropshire), Runcorn (Cheshire) and Storeton (Merseyside). St Bees (Cumbria) is also noted for its red sandstone. A much younger stone, from the Upper Greensand (Cretaceous), was mined beneath the North Downs near Reigate (Surrey) until the 1960s. Hardening on exposure, its various names include *hearthstone*.

Some softer sandstones were quarried by *guttering:* channels 4 feet (1.2 m) deep and 1 foot (300 mm) wide were cut in the surface by special picks, to divide the rock into large blocks. These were then detached from the bed with long bars and cut down to size with wedges. A great improvement on this slow and wasteful method was the steam channelling machine, introduced from the United States about 1900. However, it was less suited to harder rocks, and was therefore not adopted everywhere.

ABOVE: *Abandoned millstones and grindstones below Millstone Edge, near Hathersage, Derbyshire.*

BELOW: *Gregory's sandstone quarry and workyard at Mansfield, Nottinghamshire. Many quarries had their own masons' yard laid out nearby, with a travelling crane to serve the dressing and sawing sheds.*

A deep sandstone quarry at Higher Bebington, Storeton, Merseyside, in 1914. A steam channelling machine is at work, mounted on a tramway and using a special steel chisel to cut a channel over 10 feet (3 m) deep. Note the characteristic form of a quarry created by the machine.

ABOVE: *Granite quarrymen at Gold Diggings, Cornwall, in the 1920s. Note the two types of jumper, or 'swell jumper', used to drill shallow holes for splitting granite. Each made a different sound, and the whole was likened to a peal of bells.*

BELOW: *Plugs being hammered home between feathers to split granite at Trenoweth, Cornwall, in a method used since about 1800. The only difference is that compressed air drills are now used. Small holes of only about 3 inches (76 mm) depth are sufficient, whereas many other stones require a greater size to obtain an even break.*

A granite quarry near Penryn, late nineteenth century. Note the engineering granite being worked with hand tools and the typical timber mast cranes with wire or chain guys. Deeper quarries, notably in Scotland, used aerial ropeway cranes called 'blondins'.

GRANITE

Granite is an igneous rock, once intruded in a molten state beneath the surface, and exposed by subsequent erosion. Slow cooling formed crystals, mainly of quartz, feldspar and mica, giving granite its decorative quality, which is enhanced when polished. The feldspars are the most prominent, varying in colour from whites to reds according to their locality. Being hard and resistant, granite was much used for engineering works in the nineteenth century, especially in docks and lighthouses, while there was a great demand for it in London for bridges and public buildings, as well as for monuments, street paving and kerbs. In the early twentieth century there was a rapid decline in quarrying, brought about by imports, notably from Scandinavia, as well as by the use of cheaper alternative materials. Today few quarries are left, and most of the stone is sawn into thin slabs and polished for cladding buildings, while foreign coloured granites are also worked in the yards.

In Scotland, the 'Granite City' of Aberdeen was the centre of the largest granite district, which included Peterhead. The monumental trade was important here, with considerable exports to the United States and elsewhere before the First World War, while many new types of machinery were developed and manufactured. Quarries, such as Rubislaw, were over 400 feet (122 m) deep, but they are now disused and flooded. Off the west coast, the Ross of Mull produced a pink stone, while, further south, Dalbeattie and Creetown were once suppliers of granite for Liverpool's docks. In Cumbria, a small outcrop at Shap yields a distinct granite with large pink feldspars, valued for decorative work.

The most important English district was in Cornwall and Devon, where over three hundred quarries produced a silver-

Kemnay Quarry, in 1939. This large Scottish granite quarry dates from the 1860s, when 700 men were employed in making street setts and kerbs, mainly for the London market. In this view, finished setts and kerbs await transport in the foreground, while 'blondin' cranes are in use for hoisting stone from the depths of the quarry.

grey granite, prized for engineering work. Before the nineteenth century, the main source was from *moorstones,* boulders which lay on the moors, from Land's End to Dartmoor.

A line of grooves was laboriously chiselled out and wooden wedges were inserted; soaked in water, these expanded enough to cleave the stone. Later, iron wedges were used, driven home between plates. Plugs and feathers have been used since about 1800, inserted in a line of holes little more than 3 inches (75 mm) deep and therefore much smaller than those required for limestones and sandstones. A hand borer was struck with a heavy hammer, or a *jumper* was used. The Cornish jumper was a wrought iron rod with a sharpened cutting edge at each end, while an iron ball welded into the middle gave the instrument weight. The quarryman lifted and plunged the jumper up and down on to the same spot to make a hole. The small plugs and feathers were inserted and struck in the usual way.

The method was also used to split large quarried blocks and is still an important part of quarrying, although compressed air drills have replaced the old hand ones. The easiest direction for splitting granite is along the *cleaving* or *capping way,* which follows the horizontal lie of the rock in the ground. The *quartering way* follows the grain, or flow of crystals within the granite, and requires skill to recognise. The hardest direction is the *tough way,* which crosses this grain at right angles and needs more effort, with deeper and closer holes if the granite is to split in a straight line.

Quarries became the dominant source of granite in the nineteenth century, among the earliest being Haytor on Dartmoor, which provided stone for London Bridge (1825-31). Penryn became the major centre of the industry in Cornwall,

RIGHT: *De Lank Quarry, Bodmin Moor, in 1984. In the foreground, granite has been cut out by thermal lance, making use of rectangular jointings. It is then split into smaller blocks before being lifted out by a 10 ton electric derrick. Note the open-fronted masons' sheds (bankers) and sawing and polishing mills behind.*

BELOW: *A radial surfacer or 'dunter', worked by compressed air at Hantergantick, Cornwall. These high speed percussion machines were first introduced in the late nineteenth century. Much dust is created, hence its use in the open air and the operator's wearing of a safety helmet with respirator.*

ABOVE: *Double circular saws at work on granite slabs, with water fed into the cuts. Modern diamond-tipped saws make rapid work of all types of stone.*

BELOW: *A three-wire saw cutting granite slabs for cladding at Hantergantick Quarry, Cornwall. An abrasive is introduced into the cut, with water for cooling. Wire saws have also been applied to the winning of stone in slate and marble quarries.*

A 'Jenny Lind' radial polisher working on granite at De Lank, Cornwall, in 1978. The four high-speed carborundum heads are being moved freely across the surface.

with over one hundred quarries of sound granite within a few miles of its stone yards and shipping quays. Quarries further inland in other areas relied on mineral railways to carry their products to small ports around the coast. Widely spaced vertical and horizontal joints enabled large rectangular blocks to be quarried. Deep vertical holes were bored and packed with just enough black powder to displace a block along a horizontal joint without damaging it. Big holes 25 feet (7.6 m) deep and 6 inches (150 mm) in diameter took many days to bore by hand. Two men struck the heavy boring rod with sledge hammers, while two others were needed to give it a half turn between each blow. The introduction of compressed air drills greatly speeded the process. Since the 1960s the *thermal lance* has been employed in the larger quarries. A mixture of kerosene and oxygen is burnt at the end of a long blowpipe at nearly 3000 Celsius. This disintegrates the rock, allowing a deep channel to be cut around a large block. Once this has been obtained, secondary breaking to manageable sizes is done by plugs and feathers.

The dressing of granite is an important part of the industry. In the past, a block was first roughly *scappled* to within about 2 inches (50 mm) of its required size with a *blocking hammer*. At first, stones were shipped off in this state, but further dressing soon took place within the quarry or at a yard nearby. Chisels, such as the *point tool*, were used for roughing down the stone, while the *chop axe* obtained a finer finish. More efficient was the *patent axe*, or bush hammer, introduced from the United States. This had a number of iron plates bound together and produced a *fine axed* finish when struck at right angles to the surface. To work a large flat surface, the fast-acting pneumatic *dunter* was introduced, while pneumatic hand tools have taken over much of the more detailed carving work.

Although sawing and polishing machines have been used for other rocks, their toughest application has been for granite. There has been an increasing demand for thin polished stone cladding in the twentieth century. Sawing frames were first used, containing one or more long iron plates which were drawn mechanically back and forth across the stone. Progress with granite was about 3 inches (75 mm) per week using sand as an abrasive, but 10 inches (250 mm) per day could be achieved with chilled shot and steel plates. In the 1960s frame sawing was superseded by wire loop saws, with an endless wire travelling at high speed between two pulleys. The wire is of two strands of drawn steel, with a reverse twist every 25 feet (7.6 m) to maintain an undistorted cut. Up to 15 inches (380 mm) per hour are possible. Diamond-tipped circular saws have been used for cutting most stones for many years, and the newest large machines are fully automated, in an attempt to speed operations.

Polishing machines were introduced in the early nineteenth century. Sand and water were used on a fine axed surface beneath flat iron rings, which were turned by hand until steam power was applied in the 1830s. A great improvement, still used for smaller work, was the *Jenny Lind,* introduced from the United States in the early 1880s. Its characteristic humming was likened to the singing of Jenny Lind, the 'Swedish nightingale', who was popular at the time. To reduce a tooled or frame-sawn surface, chilled iron shot and water were applied to a rotating steel *shotting ring* at the end of the machine's radial arm. Carborundum and emery were then used to produce an *egg-shell* finish. With modern saws leaving smoother finishes, the shotting process is unnecessary. A set of high-speed carborundum wheels can be attached to the machine head instead. A final mirror finished is achieved using a felt pad with oxide of tin. At modern works automatic polishers are worked in tandem, with slabs passing beneath them without needing to stop for the polishing heads to be changed. Aberdeen was renowned for its polishing works in the nineteenth century, and huge lathes were also devised for turning and polishing columns. The largest Cornish yards had similar machines.

Part of John Freeman and Sons' granite workyard at Penryn, Cornwall, in about 1900. These columns were turned on a lathe before being further worked and erected, prior to shipment from the adjacent quay.

Underground at Llechwedd, Blaenau Ffestiniog. A long slab of slate is being hoisted on to a tramway wagon by chain, shearlegs and hand winch. Note the two methods of boring holes for reducing the size of the large block. The Ffestiniog jumper (foreground) differs from the Cornish granite jumper. The photographer's flash light is misleading; the quarrymen worked in almost complete darkness, except for the light of a few candles.

SLATE

Slate is a metamorphic rock, a fine-grained sedimentary mudstone or shale changed under great pressure during past mountain-building periods. Good-quality slate is hard and durable, with an even texture. Colour varies between districts, with blacks, greys, blues, greens, browns and purples. Slate was so compressed to form thin parallel layers, separated by cleavage lines, which allow it to be split finely to make a resistant and lightweight roofing material. Large smooth slabs can also be produced, for paving, gravestones, water cisterns and even billiard tables.

Slates were used during Roman times and again after the twelfth century. However, the greatest demand began at the end of the eighteenth century and grew thereafter, when many new houses and factories were being built. Good access to ports was an important factor in the development of the industry. This is well seen in Snowdonia, Devon and Cornwall, and particularly on the west coast of Scotland, the remotest of all, with large quarries at Ballachulish and Easdale relying entirely on sea transport.

Of the Welsh quarries, Cambrian and Ordovician slates were worked on a vast scale in Snowdonia, with concentrations at Bethesda, Llanberis, Nantlle and Blaenau Ffestiniog. The largest of the open quarries were Dinorwic at Llanberis and Penrhyn at Bethesda, the latter still working and over two hundred years old. In 1882 Dinorwic produced 87,429 tons of slates and Penrhyn 111,166 tons, when each was employing about three thousand men. They were worked by a series of large terraces or benches, hewn from the mountainside and connected by inclined railways for lowering slate blocks to the dressing sheds. Most inclines used gravity, with loaded trucks descending, while empties ascended, attached to a

Clearing rock and boring shot holes at Penrhyn Slate Quarry, Snowdonia, in about 1913. Note the thick ropes for safety when working on the quarry face.

cable around a large drum at the top. Both quarries had narrow gauge railways to small shipping docks on the Menai Straits.

Around Blaenau Ffestiniog the slate dips below ground, where it was worked out in extensive caverns, with pillars of rock left to support the roof. At Llechwedd Slate Mine, started in 1846 by J. W. Greaves, underground tramways extend for 25 miles (40 km) at different levels down to a depth of 900 feet (274 m). The nearby Oakeley Slate Mine is about twice this size. Cwmorthin was known as the 'Slaughter House' because of collapsing roofs and poor ventilation. Packhorses were the main form of transport until 1836, when the Festiniog Railway was completed to take slates to Porthmadog for shipment, allowing developments to proceed. It was connected to numerous tramways and inclines, which remain a feature of many long-closed Welsh slate quarries. The destinations of slates included Germany and the United States, and over 120,000 tons were shipped in the peak year of 1882. Other finished slates left the district on branches of the Great Western and London and North Western railways. The slate workers' village of Blaenau Ffestiniog is dominated by waste tips, for over 75 per cent of quarried rock becomes waste in the winning of slate. Rhosydd, perched 1500 feet (457 m) above sea level, is one of many remote workings which had barracks, where men stayed during the week to save a long daily walk to work.

The younger Devonian slates of south-west England were worked on a lesser scale, the major quarry being at Delabole in north Cornwall. It is now 500 feet (152 m) deep and 1.6 miles (2.6 km) around — the deepest of its type in England. 450 men were employed at the end of the nineteenth century, when the quarry was connected to the surface dressing sheds by inclined tramways and aerial rope-ways. Most finished slates were sent by wagon to nearby coves and shipped as far as Holland, although London and many

An 86 foot (26 m) ladder for inspecting the roof gives the scale of a chamber at the Oakeley Slate Mine, Blaenau Ffestiniog.

English coastal towns remained important markets. It was not until 1893 that the main-line railway reached Delabole. Cliffside quarries near Tintagel loaded their products into small vessels moored beneath, a hazardous excercise on this rough coast. There were underground slate workings at St Neot, also in Cornwall.

The greenish Ordovician slates of the southern Lake District have provided an exportable material. They are volcanic tuffs (ash), subsequently metamorphosed, and a fine bedding adds to the decorative qualities. Burlington and Broughton Moor are among the largest quarries, but there are others at Honister and around Langdale.

Although some slate rock could be levered out by taking advantage of joints and weaknesses, some blasting was required for large-scale quarrying, as today. Holes were bored with a long chisel, struck by a heavy hammer. Black powder was used to dislodge large pieces without shattering too much rock. Fallen blocks were reduced in size by plugs and feathers. Holes were made with a jumper up to 10 feet (3 m) long, but used in a similar way to the Cornish granite jumper. Suitable blocks were lifted on to a tram wagon by using a hand winch and shearlegs. At the dressing sheds a circular saw cut the rough blocks into large slabs or smaller sizes for splitting by hand for roof slates.

Splitting requires the greatest skill and is still done by hand, although some machinery has been introduced. Twelve thin slates can be produced from a slab of 2 inches (50 mm) thickness. They are then trimmed to the required size, either by hand or machine. If by hand, the dresser would hold the slate firmly on a sharpened iron edge and trim off the surplus with a steel knife called a *zax*, the sizes being gauged with a notched stick. The simplest machine at the end of the nineteenth century was a guillotine, which could be treadle or power operated. An improvement was a circular

Delabole Slate Quarry, Cornwall, in about 1900. This huge open quarry was by then over 400 feet (122 m) deep, having been worked since Elizabethan times, and served by inclines and aerial ropeways.

motion machine with two knives, doing a much more rapid job and still in use today. The main sizes of slates have interesting names, which are similar throughout the districts, although their dimensions can vary. They include (from large to small) princesses, duchesses, marchionesses, countesses, viscountesses, ladies, small ladies, doubles and singles. Rags and scantles were of irregular size and thickness, and common especially in Cornwall and Devon, where they were used mostly locally.

By the end of the nineteenth century water and steam power were used in the dressing mills, as well as for winding and pumping in the quarries and mines. Electricity was only beginning to be introduced. Large waterwheels were common in the mountains and were served by well engineered reservoirs and water courses.

Competition from other materials in the twentieth century brought a rapid decline for the slate industry. Foreign markets were lost too, particularly following the First World War. Despite the introduction of compressed air drilling and even mechanical excavators, slate production remained expensive. However, there has been a revival since the 1970s, with a demand for high-quality natural roofing materials. New uses have been found, and the Burlington Slate Company has developed an export market for slate cladding for buildings.

There are other rocks called 'slates', and the techniques of production are different. For example, the Collyweston slates of Northamptonshire are hard sandy Jurassic limestones and have been worked from small quarries or shallow mines. Extracted stones are stacked outdoors in winter, with their bedding planes vertical. They are kept wet so that frost aids their splitting into thin pieces, to be fashioned for roof 'slates'. Similar 'slates' are produced at Stonesfield (Oxfordshire).

ABOVE: *Loading Delabole slates at Port Gaverne on to the ketch 'Surprise' (42 tons, built 1879), beached at this tiny haven on the north Cornish coast. Note the women among the loaders. In contrast, Welsh ports such as Porthmadog had quays and railway sidings devoted to slate and were bustling with shipping.*

BELOW: *The slate stacking area outside the dressing mills at Llechwedd in 1895. Slates are being selected for carriage to the Festiniog Railway. Note the large waterwheels, an important source of power used wherever possible in this mountain district.*

ABOVE: *Splitting and dressing slates. On the left, the fissile nature of the slate is being exploited with a broad chisel and hammer. On the right, the slates are squared on a Greaves engine, a revolving trimmer designed by J. W. Greaves at Llechwedd in the nineteenth century. Both methods are still in use.*

BELOW: *Broughton Moor Slate Quarry, high on the fells of the southern Lake District, in 1961. Note the great extent of waste in proportion to quarry workings, some of which are entered through tunnels. Dressing mills are built on the level tops of the waste tips.*

A portable steam engine working a small stone crusher and screening plant at Gwavas Quarry, Newlyn, Cornwall, in 1906. The graded stone was sent by narrow gauge railway to the loading pier. As with similar quarries elsewhere, the coastal site enabled quarrying to develop, shipping roadstone to south-east England and the continent.

BROKEN STONE

The most important type of quarrying today is of broken stone, for roadstone, railway ballast and aggregates. Hard-wearing igneous rocks, including granites and basalts, were among the earliest sought, but other rocks have served well, such as quartzites and Carboniferous limestone. At first quarrying was for local purposes, but towards the end of the nineteenth century there was a large demand from cities for street setts, made by men working in crude shelters at the quarries, although machines had some success. As tarmacadam replaced setts roadstone became important, and as the twentieth century progressed the construction industry sought more aggregates. Originally broken by hand, rock-crushing machines and screening plant were soon introduced, becoming increasingly automated.

In the Midlands, the extensively worked granodiorites of Leicestershire and quartzites of Nuneaton were well placed by main railways and canals, but coastal sites were valued for cheap sea transport. Although Aberdeen was important for sett making, most quarries are on the west coast of Britain. Notable examples include Bonawe and Furnace in Argyll, Penmaenmawr in North Wales, and Penlee and St Keverne in Cornwall. There are many abandoned quarries and harbours in these districts, especially in Wales, among them the tiny quarry harbour at Porthgain (Dyfed) and Yr Eifl and others on the Lleyn Peninsula. Sea transport was vital for the granite quarrying industry of Guernsey and Jersey, from where large quantities of setts, kerbs and, later, roadstones were shipped to the south of England, and especially London.

Quarries take the form of an open pit, but they may be long and narrow if an igneous dyke is followed. The method of quarrying is still usually in benches, so that several faces can be worked simultaneously. Earlier hand methods have been replaced by compressed air machines for the deep drilling of shot holes. High explosives are used in blasting, to bring down large masses of shattered rock. This is aided by badly jointed and faulted rocks. *Chamber* or *heading blasts* are no longer used. A tunnel was cut over several months into the quarry face, and at the end a chamber was packed with explosives and sealed. In 1932 nearly 30 tons of explosives were placed in three chambers at Bonawe on Lock Etive, to bring down an estimated one million tons to provide three years' work for three hundred men. Large rock pieces might be broken down by further blasting. Unusually, *fire setting* (lighting a fire on top of the stone) was tried in the nineteenth century at Clee Hill in Shropshire.

Carboniferous limestone is now a major source of crushed stone, especially around Buxton and Wirksworth in Derbyshire, but also in the Pennines, South Wales and the Mendips. It is also sought for the cement and chemical industries, and for use as a flux in steel making. Many limestones have been quarried for lime and burnt in kilns, often at the quarries. Interesting old limekilns are common, both in limestone areas and around the coast, to which both stone and fuel were shipped.

A modern roadstone quarry at Machen, near Newport, South Wales, in 1978. Tipper trucks bring quarried limestone to the primary crusher (right). The rock is further graded by passing through secondary crushing and screening plant, before being stockpiled.

ABOVE: *The loading wharf of the Clee Hill Granite Company in the 1890s, when there were over 14 miles (22.5 km) of quarry tramways. Stone is being transferred to the main railway system. The stone was basalt, although called 'granite' by this Shropshire company, a practice often causing confusion to geologists when applied to other hard roadstones in Britain.*

BELOW: *Tipping waste at Burlington slate quarry, in the south Lake District, in 1945. This was hard work, not without danger, but it was a vital part of the quarrying process.*

PLACES TO VISIT

There are thousands of disused quarries throughout Britain. Old workings can be dangerous and most are on private property, so it is courteous to ask permission to visit them. On request, active quarries may allow access for interested visitors, although certain types of operation can make this difficult on grounds of safety. Several museums in quarrying districts have small collections of related material, but there is an increasing number of places devoted to the winning of stone. Intending visitors are advised to find out the times of opening before making a special journey.

Bath Stone Quarry Museum, Park Lane, Corsham, Wiltshire SN13 9LG. Open by appointment only, correspondence to: 1 High Street, Seend, Melksham, Wiltshire SN12 6NR. Telephone: Seend (038 082) 645.

Beer Quarry Caves, Quarry Lane, Beer, Seaton, Devon. Telephone: Seaton (0297) 20986.

Camden Works: Museum of Bath at Work, Julian Road, Bath, Avon BH1 2RH. Telephone: Bath (0225) 318348.

Delabole Slate Quarry, Delabole, Cornwall. Telephone: Camelford (0840) 212242.

Gloddfa Ganol Slate Mine, Blaenau Ffestiniog, Gwynedd. Telephone: Blaenau Ffestiniog (0766) 830664.

Guernsey Folk Museum, Saumarez Park, Catel, Guernsey, Channel Islands. Telephone: Guernsey (0481) 55384.

Llechwedd Slate Caverns, Blaenau Ffestiniog, Gwynedd LL41 3NB. Telephone: Blaenau Ffestiniog (0766) 830306.

National Stone Centre, Wirksworth, Derbyshire. Telephone: Matlock (0629) 3411, extension 7162. (opening 1986.)

The Slate Caverns, Carnglaze, St Neot, Liskeard, Cornwall. Telephone: Liskeard (0579) 20251.

Welsh Slate Museum, (National Museum of Wales), Gilfach Ddu, Llanberis, Caernarfon, Gwynedd LL55 4TY. Telephone: Llanberis (0286) 870630.

FURTHER READING

Bezzant, N. *Out of the Rock.* Heinemann, 1980. A history of the Bath and Portland Group.

Greenwell, A., and Elsden, J. V. *Practical Stone Quarrying.* Crosby Lockwood and Son, 1913. Valuable for the quarrying methods of the period.

Hall, C. J. *'Tanky' Elms: Bath Stone Quarryman.* C. J. Hall, 1984. A good account of a working life in the Corsham stone quarries.

Hudson, K. *The Fashionable Stone.* Adams and Dart, 1971. Describes uses and quarrying of freestones.

Lewis, M. J. T., and Denton, J. H. *Rhosydd Slate Quarry.* Cottage Press, 1974. A detailed survey and description on the surface and underground.

Lindsay, J. *A History of the North Wales Slate Industry.* David and Charles, 1974.

O'Neill, H. *Stone for Building.* Heinemann and British Stone Federation, 1965.

Perkins, J. W., Brooks, A. T., and Pearce, A. E. McR. *Bath Stone: A Quarry History.* University College, Cardiff, and Kingsmead Press, 1979.

In addition, there is the *British Regional Geology* series of the British Geological Survey, useful for geology and the location of some quarries. Quarrying museums publish booklets on their own industry.